Yamile Venticuaba Hernández
Pedro Lucas Alberto

Rural extension work to promote food security

Yamile Venticuaba Hernández
Pedro Lucas Alberto

Rural extension work to promote food security

Rural extension actions to promote food security in the locality of Ovipembe, Ondjiva Cunene Province

ScienciaScripts

Cover image: www.ingimage.com

This book is a translation from the original published under ISBN 978-3-639-68500-8.

Publisher:
Sciencia Scripts
is a trademark of
Dodo Books Indian Ocean Ltd. and OmniScriptum S.R.L publishing group

120 High Road, East Finchley, London, N2 9ED, United Kingdom
Str. Armeneasca 28/1, office 1, Chisinau MD-2012, Republic of Moldova, Europe
Managing Directors: Ieva Konstantinova, Victoria Ursu
info@omniscriptum.com

Printed at: see last page
ISBN: 978-620-8-60770-8

RURAL EXTENSION WORK TO PROMOTE FOOD SECURITY IN OVIPEMBE, ONDJIVA, CUNENE PROVINCE

Authors

MSc. Yamile Venticuaba Hernández

Eng. Pedro Lucas Alberto

2023

ACKNOWLEDGMENTS

I am grateful first of all to God for the infinite graces he has given me in my life to make this moment come true, for my health and strength, not just as a university student, but at every moment of my life.

I would like to thank my family, who have always been there for my education, as well as my teachers, especially those in the agronomy specialty, who have tirelessly taught me scientific and social knowledge, in order to put it into practice and pass it on to future generations and beyond.

We would like to extend our thanks to my tireless and patient tutor MSc.Yamile Venticuaba Hernández, who was always patient and willing to accompany this work, correcting, advising and always ensuring its success.

The institution and everyone who was able to participate directly or indirectly in its success.

Thank you very much!

DEDICATORY

I dedicate this work to my children, my parents and my siblings, for the emotional, material and spiritual support they have given me during my career and for making it possible.

To all those who, in some way or another, made my dream and the completion of this work possible.

SUMMARY

The aim of the research was to develop rural extension activities to improve food security in the town of Ovipembe, Ondjiva commune. The work took place between January and April 2023. The design was structured in phases that gave rise chronologically and systemically to the specific objectives. The research was qualitative and based on the principles of participatory action research, as described by Roa et al. (2002). The historical-logical, analytical-synthetic and structural-systemic methods were used. Research was the fundamental strategy of the investigation, which was qualitative in terms of data collection and quantitative in terms of data interpretation. The results of the research were processed using tables and graphs in order to facilitate understanding. The information was analyzed using Excel and Word for Windows version 10. A proposal was drawn up for rural extension work to promote food security in the locality of Ovipembe, Ondjiva commune.

Keywords: extensionism, food security and insecurity, proposed actions.

I. INTRODUCTION

Food and nutrition insecurity is an obstacle to sustainable development in multiple ways, and is both a cause and an effect of poverty. The number of systematically hungry people in the world has been growing since 2014, reaching 690 million people in 2019 (ten million more than in 2018), although if less acute situations are also taken into account, it can be concluded that more than 2 billion people in the world did not have regular access to safe, nutritious and sufficient food in 2019 (FAO 2020). These figures are similar to the levels of hunger and food insecurity recorded more than a decade ago, i.e. there is no improvement in the indicators, but rather a move away from the goal of eradicating hunger (SDG 2) and also little progress on various nutrition indicators

Access to affordable food that promotes health and good nutrition for a growing world population remains a major international challenge. However, we live in a world with a basic incoherence: 815 million people go hungry, out of a global population of 7.8 billion, while all the food produced today could feed almost 12 billion people and 1/3 of all food is lost or wasted. Food and nutrition insecurity is both a cause and an effect of poverty: while malnutrition perpetuates poverty and deprivation, poverty aggravates malnutrition and generates food insecurity (Magalhães, 2018).

The same source reveals that unsustainable food production models, imbalances in distribution and access, environmental degradation, resource scarcity and climate change, food waste and unsustainable consumption patterns are among the multitude of factors affecting development prospects, while at the same time being influenced by the development models pursued

According to Millennium Development Goals (MDG) (2010), the issue of food security has been a recurring discussion in several nations. In 2000, eight Millennium Development Goals (MDGs) were established to be achieved by 2015 by the signatory countries during the United Nations Millennium Declaration. The first of these goals was to end extreme poverty and hunger. Dias (2015) says that hunger and malnutrition are enemies of human development and can cause instability and conflict, reflecting not only on people's quality of life but also on the development prospects of societies and the growth potential of their countries.

In our view, these devastating effects are not only part of the problem, because residents in areas of low socio-economic status, in both urban and rural areas, are marginally deficient in micronutrients (iron or iodine, vitamin A), unable to reach their potential for physical and mental development. Clinical manifestations of these deficiencies are maternal and infant death, diminished immune response, blindness, mental retardation and anemia

Fighting hunger and significantly reducing poverty is one of the biggest challenges facing the Angolan state in the post-war period and a key factor in laying the foundations for building a more prosperous society and social justice, due to its impact not only on improving the living conditions of the Angolan population, which has been deeply weakened by the decades of armed conflict that Angola has experienced, but also on enabling more sustainable growth in the national economy (República de angola, 2009). Poverty is more severe and widespread in the countryside than in the cities, and is more pronounced in female-headed households in rural areas. According to research carried out by the FAO (2004) (Loliana Sive, Ilunga, and Massochi, 2021).

Reducing hunger requires an integrated approach that must include the following elements: public and private investments to increase agricultural productivity; better access to inputs, land, services, technologies and markets; measures to promote rural development; social protection for the most vulnerable, including strengthening their resilience to conflicts and natural disasters; and specific nutrition programs, in particular to address micronutrient deficiencies in mothers and children under five (FAO, 2014).

Agriculture in Angola before independence was the main sector of the economy and the country was a major exporter of corn and especially coffee. Family farmers, especially in the central region, were the main producers and were well integrated into the market. During the war, as mentioned earlier, a large part of the population was forced to leave their land and stop production in order to take refuge in urban centers or other safer areas of the country. Since the end of hostilities in April 2002, some of the displaced people have gradually returned to their areas of origin and started investing in agriculture again. (Republic of Angola, 2009)

According to Dias (2015), we now know that economic growth alone is not enough to accelerate the reduction of hunger and malnutrition in countries. Since a large part of the world's population lives in rural areas, rural communities are particularly important in the fight against food and nutrition insecurity. Agricultural growth with the participation of small farmers, especially women, will be more effective in reducing extreme poverty and hunger, and possibly increasing workers' incomes and creating jobs for the poor (

However, the conditions of services, communication routes and markets are still being rehabilitated. Most farmers practice traditional agriculture using hand tools for land preparation and weeding, planting local seeds left over from the previous harvest. In the provinces of Huambo and Bié, the coastal area of Benguela and in the south, in the province of Huíla, many farmers use animal traction. Some of them use tractors for land preparation and weeding; some use fertilizers and improved seed varieties (República de angola, 2009)

According to the Republic of Angola, Ministry of Agriculture and Fisheries and National Directorate of Agriculture and Livestock (2021), Within the scope of the actions of the National Department of Food Security, of the National Directorate of Agriculture and Livestock of the Ministry of Agriculture and Fisheries with the support of the FRESAN project financed by the European Union, a Vulnerability Assessment, Food and Nutrition Security - Measurement and Analysis of the Resilience Index (AVSAN/RIMA II) was carried out in the provinces of Cunene, Huila and Namibe between mid-March and the first week of May 2021.

As a result of this research, it was obtained that in the period from July to September 2021, it is estimated that around 1.32 million people (49% of the population) live in conditions of acute food insecurity, of which 38% in IPC Phase 3 (Crisis) and 12% in IPC Phase 4 (Emergency). These people face difficulties in accessing food, or are only able to meet minimum food needs with crisis and/or emergency resources and strategies. The current situation of acute food insecurity in the 17 municipalities analyzed in the three provinces is the result of:

1. The lack of rainfall has significantly reduced agricultural production, which is the main source of food for rural households, and caused animal losses due to lack of pasture and water.
2. The general rise in prices, which reduces access to food for low-income households.
3. The outbreak of the locust plague, which affected some communities.

It also notes that acute food insecurity is present in all municipalities. However, the municipalities of Cahama, Curoca and Ombadja (Cunene province), have the highest prevalence of the population in Crisis (IPC Phase 3) or Emergency (IPC Phase 4) with more than 60% of the total population in these two phases (July and September 2021).

The multiplicity of these effects calls for a path through collaborative, creative and innovative social intervention strategies, which are fundamental to working on people's potential, thus creating opportunities to raise awareness of individual and group capacities, and providing an increase in self-esteem and personal appreciation (Riches, 2002).

These are the reasons for the following:

Scientific Problem: How can rural extension work contribute to food security in the locality of Ovipembe

Hypothesis: if actions are taken to increase food production, the food security of families in Ovipembe will improve.

General objective: To develop rural extension activities to improve food security in the town of Ovipembe, Ondjiva commune

Specific Objectives

1. Characterize food insecurity conditions in the Ovipembe community.

2. Design rural extension activities to promote food security in the town of Ovipembe, Ondjiva commune.

II. LITERATURE REVIEW

2.1 Food security, concepts and dimensions

At the World Food Summit organized by the FAO in Rome in 1996, the Heads of State and Government affirmed "the right of everyone to have access to safe and nutritious food, in line with the right to adequate food and the fundamental right of everyone to be free from hunger", and committed themselves to implementing this right in its entirety and to achieving it gradually, with the aim of guaranteeing food security for all by the year 2015 (FAO, 1996).

According to Ferrão (2002), the food security problems experienced by the poorest countries are of the same nature, although considerably more aggravated and diversified, and therefore much more of a priority, than those currently worrying the more developed countries. Those that refer to food safety, which today are rightly of greater concern, because those that refer to food security have already been targeted, also affect poor countries, but when hunger is rife in these countries and food shortages are the norm, food security takes priority, because survival is at stake.

In turn, the FAO (1996) says that food security exists when people have physical, social and economic access at all times to safe, nutritious and sufficient food to meet their dietary needs and food preferences in order to lead an active and healthy life.

According to Simon (2009) cited in Fortes (2012), there are four dimensions to food security:

Availability, in the definition equates to the term sufficient, the amount of food that is present in a country or region, thanks to any form of domestic production, imports, food reserves or food aid.

Access, in the definition, refers to physical, social and economic access. Physical access means having a place where you can buy food; economic access means food being available where people need it and families having the financial capacity to regularly buy sufficient quantities of food to meet their needs; social or sociocultural access means situations in which the product is physically available to the consumer, who has the economic resources to buy it, but there may be sociocultural barriers that limit access to it, especially for certain groups of the population, for example, for social or gender reasons.

Use, in the definition, refers to safe and nutritious food that meets your dietary needs. Here, the selection of food products, their preservation and preparation, as well as the intake and absorption of nutrients are important. The food must be safe and of good quality. The use of food is also related to drinking water, hygienic conditions and sanitary assistance. So this dimension not only refers to nutrition, but also to other elements relating to the use, conservation, treatment and preparation food products. All of this shows a close relationship between nutrition and food security; you can't have food security without proper nutrition.

Stability, in the definition, includes the expression "at all times". Stability applies to the first three dimensions of food security mentioned above. Food security is a situation that doesn't just have to happen at one moment, on one day or in one season; it has to be permanent and sustainable.

The same researcher defines chronic and transitory food insecurity based on the stability dimension

- Chronic food insecurity - is the long-term or persistent inability to meet minimum food needs.
- Transitory food insecurity - is a short-term or temporary food deficit

Food security is difficult to measure, and so food insecurity is usually measured, assessed or analyzed with the aim of determining the factors that have lost their way and caused situations or will cause them in the future, and deciding on corrective measures.

In this case, the FAO defines food insecurity as a situation in which some people do not have access to sufficient quantities of safe and nutritious food, and therefore do not consume the food they need to grow normally and lead an active and healthy life.

2.2 Food production and hunger

Food is considered a source of survival for human beings. In order to eat well, it is essential to have access to quality and quantity to nourish the main organs of the human body. In this sense, the production of food and its availability are considered essential for adequate nutrition to take place. If not, hunger will prevail.

According to the FAO (2014), food security is a necessary condition for ensuring nutritional security, incorporating four fundamental dimensions (Figure 1): food availability, access to food, utilization of food and nutrients, and stability, which is a transversal dimension, guaranteeing the fulfillment of the other three over time.

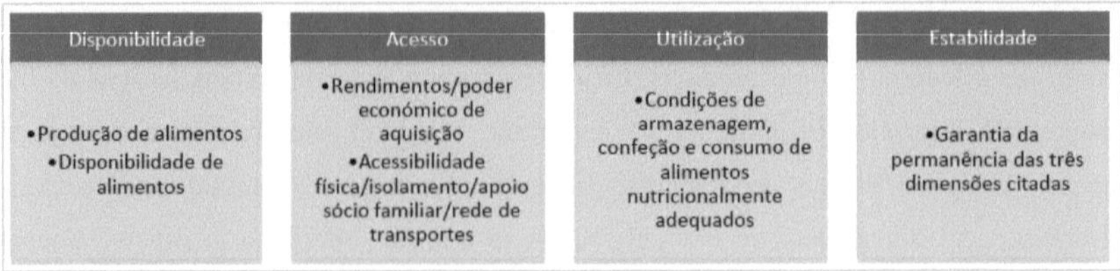

Figure 1. Dimensions of food security FAO (2014).

Castro (1968) states that the human organism is a machine very similar to manufactured machinery. Any work it does is always the result of a transformation of energy, which takes place in the intimacy of its complicated mechanism. The fuel that the human machine uses is food, which, when burned in the human engine, provides the energy necessary for its vital functioning, just like stone coal or oil when burned in combustion machines.

The same author also emphasizes that, in order to be rational, food must be sufficient and complete, i.e. it must provide the body with the total energy it needs and the total of the various substances, as essential principles nutrition - indispensable to its material balance. According to Scialabba (2011), food availability is defined as a sufficient quantity of food of an appropriate quality for access, provided through domestic production or imports, including food aid.

For the author, the availability of food is closely linked to the availability and use of natural, human and economic resources in an appropriate way for well-being.

In contrast to the availability of food, the precariousness of quantity of food available for ingestion, or its low quality, has a negative effect - it is considered hunger. It is agreed that the concept of hunger itself has a variety of meanings. As Jonsson (1981) argues, there is a distinction between the name and what is called. Moreover, translated into other languages, "hunger" can mean many different things. It is therefore important to define the use of the word. The author points out that the individual may or may not recognize hunger. Some people affected by hunger feel hungry, others don't, because they adapt to a lower food intake, thus reducing physical activity. In other cases, a lack of specific nutrients can cause hunger that is not felt by the individual (malnutrition or hidden hunger).

Jonsson (1981) argues that hunger is closely related to poverty. Indeed, poverty can often be measured by the prevalence of hunger. But hunger is not equivalent to poverty, since in some countries people are poor but not hungry. At the same time, poverty is much more than hunger, given that in some countries people are not only hungry, but also deprived of most of their human rights. It also states that hunger is a specific issue that only the human body reacts to, insofar as the food ingested is not enough.

For Castro (1968), hunger is a threat to the survival of the human race. He states with determination that the fight against hunger is a kind of cold war that threatens to freeze all the vitality of our civilization if we are not able to win it with determination

For every human being, vegetables represent the only source of nutritious sustenance to produce energy, regulate bodily functions, nourish themselves and live. From an economic and social point of view, they are important because they are a source of food and employment, because of the demand for food in all social strata and because of their high value and industrialization in local, regional and national markets. In terms of food, they are considered important for the human diet, as they provide the human body with vitamins, minerals, carbohydrates and fibers that are essential for the normal development of the individual and the prevention of many diseases (Bettiol et al., 2014).

2.3 General elements of traditional agriculture in Angola

Ecological constraints (climate, etc.) and people's response to these constraints in order to better adapt to the environment will result in the distribution of the population in Angola. According to Simões (2019), with almost all of its population essentially rural, of different ethnicities and, as such, with different socio-economic structures, Angola cannot be attributed to a single rural world, but to several. these rural areas are naturally associated with different types of economy. The peoples of the south-west, whose economy is geared towards hunting and gathering, with almost no agricultural production; the agricultural economy of the peoples in the north, with almost no livestock; the pastoral economy of the peoples on the coast-south and south; and the agro-pastoral economy of the peoples in the center.

Based on the issues already addressed in the previous sections, we have tried to make a summary analysis of traditional Angolan agriculture. At this point, a brief comparison was made between the two sectors that characterize Angolan agriculture: the business sector and the traditional sector. Subsequently, the territory was divided into large agricultural regions, according to the population density of the areas and the predominant type of activity practiced, agriculture or agro-pastoralism (INE, 2018).

Rocha, Mora, and Romero (2016) in their published article state that the biggest problem of poor agricultural production is that most rural communities lack food security because they are unaware of technical aspects in the management of their crops, such as poor soil conservation practices (derived from low soil fertility, poor drainage); Deforestation, irregular rainfall and soil erosion, cultural, traditional aspects of farming, lack of knowledge, technical assistance in planting other crops, lack of resources for purchasing inputs, no alternatives for production and diversification in agriculture, production systems, all these factors limit production, putting their food security at risk.

The available resources are not used well, as they are used indiscriminately and in an irrational and unsustainable way. Most rural communities are vulnerable due to poor agricultural production (MINADER, 2022). Furthermore, most research into traditional agriculture suggests that small-scale systems are wisely productive, biologically regenerative, energy efficient and socially just (Altieri and Nicholls, 2009). They constitute diversified production systems, which are based on the search for different returns over the course of the year, ensure family self-sufficiency and reduce dependence on external inputs.

2.4 Family and business farming

For Chivala (2019), family farming in developed countries is completely different to that in developing countries, such as Angola. In the former, this type of agriculture is integrated into the market and is committed to producing marketable products. Family farming is therefore usually referred to as traditional or peasant farming. This term is used to identify the social class that lives in the countryside and essentially does subsistence farming, only selling surpluses in local markets. The inclusion of this social group is important for characterizing Angolan agriculture today.

Peasant, this word is defined as "Rural family households with access to land as a means of livelihood, which mainly use family labour in agricultural production, integrated into a broader economic system, but which is fundamentally characterized by partial integration into markets that tend to function with a high degree of imperfection" (Chivala, 2019).

In Angola, in general, the quantities produced by Family Farming are higher than those produced by Corporate Agricultural Companies. According to MINAGRIF, the main crops grown for the 2017-2018 campaign are shown in the table below.

Table 1. Production, area and productivity of the main crops in the 2017/2018 campaign

Culture	Production (t)	Area (ha)	Productivity (Kg//ha^{-1})
Cereals	**2.885.228**	**5.171.044**	**999**
Corn	2.765.316	2.639.535	1.097
Massango/Massambala	101.214	384 849	297
Rice	9.591	8.417	1.256
Roots and Tubers	**10.876.856**	**1.186.221**	**11.963**
Cassava	8.730.517	944.152	12.954
Queen potato	458.217	63.845	15.948
Sweet potatoes	1.688.122	178.225	9.790
Pulses and oilseeds	**571.002**	**1.046.376**	**565**
Peanuts	212.089	337.971	665
Beans	233.647	672.526	493
Soy	35.266	35.879	1.001
Fruit trees	**5.211.596**	**261.498**	**22.319**
Banana	3.954.036	162.154	28.288
Citrus fruit	417.255	38.412	10.988
Sleeves	230.454	22.631	12.531
Vegetables	**1.900.006**	**132.721**	**14.684**

Source: MINAGRIF, (2018)

There is also another term advocated by some authors, such as Avillez (1973), designating this type of agriculture as traditional, defining it as agriculture that has reached a certain long-term equilibrium in terms of the use of available factors of production and the investments needed to increase the supply of these factors.

For this same source, entrepreneurial agriculture has a broader meaning compared to traditional agriculture. It applies to both developed and developing countries. The aim of this type of agriculture is "to ensure that the company's net results are maximized, and its economic results are essentially related to the return on the equity invested and the entrepreneurial capacity demonstrated".

2.5 Condition for the development of sustainable agriculture

For Ruíz (2009), the crisis in agriculture has two dimensions: socio-economic and ecological. The negative effects have to be assessed in these two dimensions. Low-income farmers are economically insolvent due to the high costs of modern agricultural technologies, which consume any increase in profits.)

The concept of sustainable agriculture has been adopted by the Food and Agriculture Organization of the United Nations (FAO, 1998). Sustainable agriculture is agriculture that achieves management capable of satisfying human needs while maintaining or improving the quality of the environment and conserving natural resources. It is one in which a balance is sought between economic growth, equity and environmental sustainability through a regulatory mechanism that is social and effective participation.

There are many definitions of agriculture, including Altieri (1997) and FAO (2003), who state that it must take into account the four pillars of sustainability in agriculture: ecological sustainability, economic sustainability, institutional sustainability and socio-cultural sustainability.

The balance of microorganisms-plants-animals-humans-environment, in its purest forms, still exists in certain regions of the third world, in underdeveloped communities in Asia, Africa and Latin America. In Africa, for example, a large part of the population is rural, very poor and dependent on natural resources and traditional production systems, which are aimed at the self-sustainability of communities. In America, in the Amazon region and the highlands of the Andes, there are also communities that exemplify these human-nature interactions; they maintain the ethnobotanical knowledge of their ancestors about the useful value of germplasm and the traditional methods for its exploitation, all in constant evolution to ensure the continuity of life (FAO, 2015).

2.6 Sustainable development in agricultural systems

Although it is known that development is a term related to growth, stability and modernization, it is necessary to recognize that it is a very complex concept. It not only has an economic meaning or material growth, but also pursues the full realization of the human being. More recent visions of sustainable development link it to a process of sustained and equitable improvement in people's quality of life, based on appropriate conservation and environmental protection measures (Espinosa, 2001).

Rodríguez (2004) conceives of sustainable development as a process comprising three dimensions: economic sustainability, ecological sustainability and technical sustainability, and explains the need to have an appropriate strategy in relation to the environment and, at the same time, ecological sustainability (conservation of the basic functions of the ecosystem, prioritizing its renewable use, recycling processes and the conservation of biodiversity).

When talking about Agricultural Development, Sarandón et al. (2006) indicate that it is a process through which greater commercial production is achieved with the available and potential agricultural and livestock resources, through the expansion of the frontier, diversification of production, technological change, increased productivity, changes in production infrastructure, among others. Like any process, it involves systematic changes coming from a socio-cultural environment.

According to Ortiz (2003), sustainable production systems are mostly characterized by maintaining practices adapted to the biophysical environment, a high diversity of crops and a low dependence on external inputs. However, these systems are mainly threatened by environmental deterioration and pressure from external technologies.

Generally speaking, as stated by Lopez (2007), a sustainable agricultural system is based on specific and precise production standards whose aim is to achieve optimal agro-ecosystems that are economically profitable, socially participatory and ecologically balanced. This highlights the fact that the needs for the development of sustainable agriculture are not only ecological or technical, but also socio-cultural, economic and political.

2.7 Main constraints, objectives and challenges in Angolan agriculture

The agricultural sector is important not only in terms of promoting self-sufficiency and food security, but also in terms of providing raw materials for the processing industry and creating jobs, one of the crucial problems in rural areas, from the perspective of diversifying the economy. For the agricultural sector to grow sustainably, agriculture needs to evolve beyond the mere subsistence level. According to CEIC/UCAN (2021), promoting agriculture is fundamental when thinking about establishing a new economic growth model that is intended to be sustainable and, above all, less dependent on oil

The same literature adds that, as such, the main problems (obstacles) should be identified and solutions sought. The majority of the cultivated area is under the responsibility of family farms, with an average area per farm of 2.32 ha, using low-efficiency production techniques, being not very productive, making it difficult to create production surpluses because small family producers have little incentive to produce beyond the family's subsistence levels.

The Ministry of Agriculture's report on the results of the 2018/19 campaign mentions the low number of tractors, and that they are only used in the business sector. The animal-drawn plough is used in both the family and business sectors, but with an average per farm of no more than 0.2 and 0.3 tools in the family and business sectors, respectively. Some of the constraints on the development of Angolan agriculture are related to the low level of human capital and knowledge, the availability of financial resources and the lack of incentives for producers and technicians. The low level of human capital and knowledge involves aspects related to:

- Low yields due to the low level of technology and difficulties in accessing quality agricultural inputs in the necessary quantities and at the right time, such as quality seeds suitable for the region, tools and utensils suitable for working the land, animal and motorized traction, fertilizers, means of preventing and combating plant and animal pests and diseases;
- Lack of an effective rural extension service that can contribute to technological progress;
- Research is not very advanced, with a lack of investment in programs and laboratories;
- Cultural barriers and a lack of knowledge on the part of farmers and technicians in technical, management, market and other terms, necessary for the projection of modern agriculture;
- Poor business management;
- High production costs associated with importing inputs;
- Inexperience in operating in a market economy system on the part of the various authors, including the government;
- Poor condition of roads, especially secondary and tertiary roads, and bridges, which hinders not only access to markets for the disposal of products, constituting a major obstacle to the marketing of agricultural and livestock products, but also access to the market for inputs;
- Lack or deterioration of other infrastructures, such as facilities for the different services, product storage and seed conservation;

The lack of incentives for producers, service providers and technicians has consequences

- Rural exodus, with the result that new generations are increasingly disconnected from the land and unfamiliar with the techniques of working the land, making it difficult for them to return to the countryside;
- Growing feminization of agricultural activity, due to the exodus of young people to urban centers;
- There are no rural financial services or they are very weak when they do exist;
- Possible land conflicts between family farmers and businessmen due to the effects of the latter's presence in unused areas;
- The population's low standard of living, which limits expectations and demand for consumer goods;
- Unattractive environment for business investment.

Among the various specific objectives defined, we can highlight the formulation of an agricultural policy, the development of agricultural research capacities, the implementation of a credit line to support associations, cooperatives, small and medium-sized producers, the rehabilitation of infrastructure to support productive activity, the promotion of trade in rural areas, the integrated promotion of strategic sectors, the diversification of the economy and the promotion of public-private partnerships. Priority areas for intervention in terms of agricultural policy for the development of agriculture include the promotion of rural trade, technical assistance and agricultural extension services, agricultural credit and scientific research (CEIC/UCAN (2019).

In order to combat poverty and the lack of food security, African organizations have taken measures and made efforts in various countries, including Angola. In July 2002, the African Ministers of Agriculture approved the Detailed Program for the Development of African Agriculture (PDDAA), a program aimed at restoring agricultural growth, rural development and food security (MINADER, 2004). Five pillars of the program are stated:

1. Expansion of the area covered by sustainable land management systems and safe control of water resources;
2. Improving rural infrastructure and trade-related capacities for better access to markets;
3. Increasing food supply and reducing hunger;
4. Development of agricultural research, dissemination and adoption of technologies to support long-term productivity growth;
5. Sustainable development of livestock, fisheries and forest resources

The participatory development of rural populations is part of the food security strategy. Policies and programmes to combat hunger and poverty include credit programmes, training and diversification of institutional solutions (public agencies, community groups, participatory organizations), support for decentralization and public sector reforms. There is also the promotion of the development of local capacities, involving relations between community groups, NGOs and companies, through the financing of small investments (Angola, 2005).

2.8. Need for training for producers in rural areas

Rural development depends much more on the adequate training of farmers than on the abundance of their resources; much more on intellectual inputs than on material inputs; much more on how to do things and what to do them with. A large part of farmers' problems can be solved by themselves, provided they receive training aimed at producing economic results and not just carrying out activities; training that is more committed to solving problems than to problematizing solutions. This makes it possible to overcome, to a large extent, the scarcity of material and financial resources for rural producers and the lack of government support (Lacki and Marcenaro, 2005).

For these authors, a large part of the increase in production and competitiveness achieved in developed countries is due to the professionalization processes of their producers and agro-entrepreneurs, who have been constantly incorporating new production and communication technologies that have enabled them to maintain competitive advantages in the face of market globalization.

In this way, the high professional levels and skills of their human resources are the main factor that allows these countries to remain integrated into world markets and constantly adapt to the changes that are taking place in these markets, without creating dislocations in their internal economies (Münch, 2005). These investigations generally lack the participation of the main players in the production process and their up-to-date training in the new advances in science and technology in the agro-ecological field (Altieri and Nicholls, 2009).

The participation of farmers in agricultural research implies a systematic dialog between farmers and scientists, aimed at solving problems related to the impact of research. They have different needs, cosmo visions, knowledge systems, methods and tools, but they can result in more productive, stable, equitable and sustainable agricultural systems (Bellón, 2002).

The need to solve these problems and the results of various consultations and diagnoses carried out by different institutions highlight the fundamental importance of implementing solid distance learning and retraining programs that are sustainable over time and take advantage of the new information and communication technologies. It is necessary to rethink and reorient training programs in order to develop all the skills required to integrate with new trends among trainers, teachers, producers and other actors linked to rural development, and to promote the generation and application of new knowledge for the sustainable development of the agricultural and livestock sector (Ramírez and Rodríguez, 2002).

The agriculture of the future must be sustainable and highly productive if it is to produce food for a growing human population. Organic agriculture has succeeded in demonstrating that it is possible to produce enough food without damaging our planet's natural resources. When we stop using agrochemicals, we stop polluting the soil, water and air. We leave our children a planet that is in a position to feed and sustain them, and this is what we call sustainable agriculture (Vázquez, 2010).

To achieve an increase in productivity, improved health and product quality, one of the essential factors is proper fertilizer management. This practice boils down to satisfying the plant's nutritional needs at all stages of development through techniques that provide greater optimization in the use of water, fertilizers, labour and other inputs, culminating in lower nutrient losses through erosion, leaching and volatilization (Souza et al., 2003).

In recent years, vegetable production has become a means of obtaining high economic returns and is a way of improving the diet of many urban and rural dwellers. It is closely related to agricultural and rural development, is a source of employment for labor that would otherwise be underused and helps to maintain good nutritional levels (Saavedra, 2013).

2.9 Rural extension, brief considerations

In our opinion, Angola has the conditions and potential to aspire to higher agricultural productivity and production volumes. One of the main obstacles has been the difficulty in accessing new production technologies, which means that resting is still very common and that little animal or mechanical traction equipment is used. In this sense, the Angolan government has implemented various programs in an attempt to modernize agriculture in recent years. However, this process doesn't just depend on access to technology, so a process of transition and learning is necessary. The low level of training and qualification of technicians and rural producers in the use of these technologies has become a new obstacle to the sector's growth.

According to Andrade, Santos and Chagas (2012), Technical Assistance and Rural Extension (ATER) were set up in Brazil to contribute to the project of developing capitalism in the countryside. Between the 1940s and 1960s, ATER organizations were created with the aim of implementing the "Green Revolution". The state established a set of actions to create the conditions, including technical assistance and rural credit.

In order to convince the peasants to abandon the agricultural practices they had been practicing and take on the "technological package", courses and schools were set up to train technicians in agro-livestock farming. The Technical Schools have the task of training these technicians, who in turn will have the role of passing on the technical procedures of this new technology to the peasants. However, the technicians' tasks go beyond technicality; they are also a link in the "transmission belt" of the ideology of the ruling class (Freire et al., 2015).

The introduction of rural extension organizations in African countries took place a little later, with most being created between the 60s and 70s. In most third world countries, the general creation of rural extension organizations was carried out through external assistance, particularly from the United States. Since at the time of independence few Third World countries had agricultural schools or universities, in almost all of them rural extension was assigned to the Ministry of Agriculture and not to an agricultural school as in the United States (Swanson, 1991).

Rural extension is characterized by having four fundamental roles, according to Burton (1991), which are described as follows

Knowledge and skills: Although farmers already have a lot of knowledge about their environment and production system, extension can bring other knowledge and information that they don't have.

Technical advice and information: On production activities in production areas to improve their production. Extension provides information to assist producers in making decisions and, in general, to enable them to act. For example, information on prices and markets, availability of credit and inputs.

Producer organization: Producers also need some form of organization, both to represent their interests and to give them a means for collective action. Extension should therefore be concerned with helping them to establish local producer structures and organizations. This help should only be given in consultation with the farmers and with their efforts. , the organizations will not be sustainable.

Motivation and self-confidence: One of the main obstacles to development that many farmers face is isolation and a feeling that there is little that can be done to change their lives.

It is important to encourage farmers to take the initiative and involve them in extension activities and that they can do things for themselves, that they can make decisions and that they have the capacity to solve many of their problems.

2.9.1 Agricultural extensionism

Jacobsen (1994) reveals that extension is an educational medium between intercommunicated people and at the same time "accompanies" producers in the application of knowledge and technological advances, taking into account their own knowledge and skills in their work. Landini and Bianqui (2014) agree that, in itself, agricultural extension is a privileged strategy for tackling rural poverty, due to its potential to promote rural development processes and socio-productive transformation.

However, training is a teaching-learning process aimed at incorporating new knowledge, attitudes and skills to contribute to finding solutions to socio-economic, environmental or community problems. It is developed through appropriate communication, using different languages (oral, written, mime, symbols) as an instrument, which means that training has an essential dimension; it is not just a practice of talking, but of using the language that is necessary or most appropriate (Instituto Nacional Tecnológico-INATEC)

According to Engel (2000), extensionism can have positive or negative consequences, depending on the objectives pursued, the clientele and the assessment of success. In particular, it has failed to convince producers to adopt a particular recommendation, when in reality the technology being promoted was not the right one. But it is important to produce a change that fundamentally reorients the conceptualization, approaches and methodologies of extensionism. It is not enough for extension to be a vehicle for transmitting research results to agricultural producers, but it must be positioned as an instrument for strengthening rural communities' capacity for self-learning and permanent innovation for competitiveness and sustainability.

In this regard, Beltrán (2016) emphasizes the need to create the necessary conditions for the social and productive development of rural inhabitants, as well as spaces for the participation and inclusion of different social and political actors in the planning of territorial development, within the framework of a peace agreement with institutional support, thus adding environmental sustainability, through friendly production and the protection of natural resources.

There are principles that govern the process of extesionism, some of which are provided by Vega-Dorantes and López-Barbosa (2014).

- Promotes the development of the rural population through education.
- Oriented towards education, supporting people to help themselves.
- It reaches everyone. It doesn't discriminate between political parties, races or religions.
- It is based on solidly proven knowledge.
- It starts with the people according to their educational, economic and cultural level. Plans cannot be drawn up without prior education.
- It promotes the participation of volunteer leaders and encourages their training.
- It works with all family members, considering them as the basic social group for rural improvement.
- Encourage the active participation of the rural population for their own well-being.
- It encourages the development of the individual, respecting their freedom of decision and action.
- It's a two-way street: bringing scientific information to people in the field and bringing the community's problems to scientific institutes to find solutions.
- Extension programs go from the bottom up. Extension coordinates with established organizations.
- The evaluation and investigation of programs, methods and procedures is an ongoing extension process.
- It acts in harmony with agrarian policy at all levels.

However, they should be integrated so that a more objective context can be used to promote better extension, with the aim of gaining a real insight into the problem of the state's agricultural and livestock activities and the resources available to tackle it.

According to Rendón et al. (2015), rural extension in Mexico has become a dispersed and disjointed effort through training and technical assistance programs aimed at productivity as a goal and at the producer only as an element for adopting technology. This process has neglected the aim of sustainably improving the quality of life of the rural population, which requires the participation of the different actors and recognizing the authentic needs of the population and their production systems; a process in which research is required.

Huesca et al. (2015) point out that local knowledge and the recovery of experiences in the communities themselves through greater interaction between the technician and the small producer are fundamental parts of agricultural extensionism, and have a greater impact if the technical-experimental work is carried out in a participatory manner and in the communities themselves, a situation demonstrated by this research.

III. MATERIALS AND METHODS

3.1 Location

The research took place between January and April 2023, in the locality of Ovipembe, Ondjiva Commune in the Cuanhama municipality, Cunene province, which is located in the southwest of the country and borders the Republic of Namibia.

3.2 Population and sample

The population consisted of 11 peasant families and workers from the town of Ovipembe, and the sample consisted of 5 families, which represents 45% of the population. Several factors were taken into account as selection criteria, such as distance and travel conditions, and possession of land for food production and economic support

3.3 Methodology used

The research was qualitative and based on the principles of participatory action research described by Roa et al. (2002). The methods used were

Hypothetical-deductive**:** used to organize knowledge and demonstrate the objectives

Synthetic analytical: needed to collect data and information on the study region, and structural systemic: used to shape the proposal.

The research was carried out in (2) two stages.

3.3.1 Research stages

Stage I: Characterization of food insecurity conditions in the Ovipembe community

In order to ensure that this step was carried out, different techniques were used to collect data in the research area.

Interview: this was carried out with officials from the Ministry of Agriculture and Public Health, with the aim of obtaining information related to the characteristics and efficiency of the production process on peasant farms, levels of training in terms of agriculture, and the health and food security situation.

Survey: this was carried out among local farmers in order to find out their criteria regarding the productivity, consumption and marketing of the products from the fields in the study area, as well as their personal perception of the availability of food Annex 1)

Observation: it was carried out in the different areas of crop and animal production, as well as in determining manifestations of malnutrition (annexes 3, 4 and 5).

Indicators that show the objectivity of food safety conditions in the study area were evaluated.

Indicators evaluated

- Maintain and use land by peasants
- Levels and destination of dryland food production, horticulture and livestock.
- Existence and manifestations of diseases associated with malnutrition.
- Technical training in the region

Statistical Analysis

The information collected was processed by coding the survey questions, making graphs, diagrams and data matrices. The variables evaluated are shown in the results and discussion of the work using tables and graphs, processed with Microsoft Excel version 2010 tools

Stage II: Designing a proposal for rural extension activities to promote food security in the locality of Ovipembe, Ondjiva commune

The aim of this proposal is to develop technical skills in families with food insecurity and shortcomings in the operation of food production systems, in the locality of Ovipembe, Ondjiva commune, under the principles of care and protection of the environment.

The methodology used is based on the principles proposed by Jiménez (2015) (FAO Consultant) (Annex 2). Its general aim is to provide a technical-methodological tool for developing training for farmers and family members on agricultural and livestock issues, using participatory techniques that facilitate the participants' learning.

IV. RESULTS AND DISCUSSION

4.1 Results of the survey of farmers in the locality of Ovipembe, Ondjiva commune

Characterization of food insecurity conditions in the Ovipembe community

The sample under study has an average age of 30 - 50 years. These figures suggest that the farmers are adults, with responsibility for creating a source of income to support their families. The level of education of the heads of household also varies from one to another, with one not being able to read and four being able to, three of whom have primary education and one secondary education, showing a low level of education, which points to low knowledge of the production and marketing processes and their alternatives for finding and preserving food. In terms of gender, 25% are female, which still reflects a panorama of inequalities that, in some way, condition women's economic affirmation, despite their central role in carrying out different agricultural production activities, marketing products and caring for children.

INE (2020) in its Thematic Report on Gender reflects that 56.1% of households headed by women suffer from food insufficiency, while in households headed by men, the percentage is 55.6%. Although the data shows that food insufficiency is slightly higher in female-headed households, we should not lose sight of the fact that the difference is not marked, which shows that it is a generalized situation.

In other respects, it is worth highlighting the shortcomings in terms of irrigation, which depends on the rain, and has had difficulties in sustaining their families, because sometimes the rain is late, and if it is to fall, it has been too much, dragging the seeds through the soil. The consumption of water from the chimpacas, which their livestock depends on, makes it very difficult for the community to control diseases, given that it is still, untreated water, which requires a lot of care when consuming it

On the subject of safe food, according to the 1996 Rome Declaration, safe food occurs when all people have permanent physical, social and economic access to safe, nutritious food in sufficient quantity to meet their needs . But due to the lack of financial conditions on the part of the communities, the diet is not complete and few people eat enough nutrients, which is confirmed by the fact that 100% the sample refer to a lack of conditions, according to them the community does not receive any support from a non-governmental program that could support hunger and other needs, and few people eat eggs, vegetables, bread, oil and much more.

There is little agricultural production, since they lack the technical training to produce other crops, such as potatoes, yams, beans, cassava and fruit, which they say are not part of their culture and due to the soil and climate conditions. As for preparing the soil, it's all done by hand and cultivation depends on donkeys and cattle with the avenca plough. They've also had difficulty acquiring seed, fertilizer and organic fertilizer, i.e. for their vegetable gardens, and lack support from the Ministry of Agriculture.

The schools that exist there (primary and secondary) unfortunately don't have any training in agriculture. As far as cattle are concerned, their production is low and they are not sold. As in other parts of the world, meat should only be killed if there is a high-profile festival, such as an ephic or even a death. Also, the low production of pigs is due to a lack of food to sustain them. The animals' food depends on nature, in the case of grass and leaves. There are also families who don't have animals and have sometimes been in great need, due to the lack of a source of protein.

It is worth noting that a similar situation was investigated by Mendonça et al. (2015), who pointed out that despite advances in differentiated policies for family farming and social policies, a significant number of people in rural areas remain unreached by public policies, creating a hard core of rural poverty. Family farmers in a situation of food insecurity, without access to technical assistance; without documentation; living in places that are distant or isolated from public authorities; with poor conditions for generating surpluses, lacking water and energy infrastructure, with major difficulties in accessing credit and marketing channels.

Maintenance and use of land by peasants

The results of the survey show that 100% of the farmers own their own land, which reduces the loss of money from renting areas to carry out agricultural and livestock activities. Of the rainfed land, 60% claim to have 5 ha, 20% 6 ha and 7% 7 ha respectively; 100% of them work manually; of those who have water for irrigation, 60% have 5 ha and 40% 4 ha. The drawback is that only 20% use hoses for irrigation and 80% do it manually.

Table 2. Areas farmed by peasants

Tranche	Area (m^2 or ha)	Form of Operation	Paid Income (in Kz)
Sequeiros	60 % 5 20 % 6 20 % 7 ha	100% Manual	Own land
Watering	60 % 5 40 % 4ha	20% 80% Manual	Own land

These results reflect that the practice of agriculture in the territory is precarious, if we take into account that all producers own land, but do not have technologies that make it easier to work and make the most of the water and humidity in the soil.

These are shortcomings in guaranteeing an efficient production process that ensures the productivity of the crops being farmed, which can be compared, even when there are differences between them, with the statistical results of INE (2020), which show that 51.2% of respondents said they had land titles. The lack of titling, the licensing process and the attribution or recognition of land rights has been pointed out as one of the biggest sources of land conflicts and fragility in safeguarding land rights, especially property rights. More than 30% of men and women said they had inherited their plot of land, while of women and 24.1% of men said they had been given it free of charge.

The same source says that land is an important resource. In economic theories it is considered capital, as it can generate income and revenue. Either through the work applied or through the transfer of its use. In many economies, land is considered a strategic resource, as it serves multiple purposes, such as: housing, agriculture, grazing, resource extraction, study and research, etc. More than 85 % of people use their land for agriculture, suggesting that the majority use the land for food production, while around 7 % do not use their land. The question is the size, scale and quality of this production. This shows that land is a fundamental resource for household income and the living conditions of its members.

The food and nutritional security of the entire population, sustainable rural development, social inclusion and the fight against hunger and poverty are, directly or indirectly, related to the conservation and use of biodiversity resources. Today around 75% of the world's poorest - 1.2 billion people - live in rural areas and depend on agriculture for their livelihoods (Santilli, 2009).

Levels and destination of food production

The table shows the total rainfed production of the five families, which amounts to 1457.6 Kg, of which 1196.308 Kg are destined for consumption, which represents 82%, and 260.652 Kg, which corresponds to 18%. The average sale price is 2583 Kg for a total income of 64147.5062 Kg. These results show that the largest amount of harvested produce is destined for the families' self-consumption, with a very small percentage being sold, which will guarantee an increase in the money needed to buy other basic foodstuffs such as sugar and salt, as well as meeting other clothing and footwear needs. These results can be analyzed according to Cultrera (2008).

Table 3. Dryland yields

PRODUCTS	USE						
	PRODUCTION	SELF-CONSUMPTION		FOR SALE		SALE PRICE	TOTAL SALES REVENUE
	(Kg)	Kg	%	Kg	%	Kz	Kz
TOTAL	1457.6	1196.308	82	260.652	18	2583	64147.5062

Small farmers, whether they belong to traditional populations or not, are the world's greatest promoters and holders of agricultural diversity and have a strong connection to their way of life. Their crops are based on subsistence production, with the possibility of selling surpluses, family labor, simple technology and, above all, growing several different species in the same area. These practices guarantee food and nutritional security, while at the same time ensuring planting against abiotic and biotic stresses, such as climate change and attacks by pests and diseases (Cultrera, 2008).

Figure 2 shows the diversity of crops produced under rainfed conditions. In line with the previous table, it can be seen that the largest amount of production is destined for self-consumption, with 81%, 84%, 85% and 86% of watermelon, maize, millet and ginger, respectively, being the most consumed, while the most sold are 23% of pumpkin and cucumber, 25% of beans and 29% of quizaca. If we take this into account, the production of grains and tubers is not very varied and the percentage of sales is not high enough to make up for the deficiencies with products from stores and stores

This coincides with what the scientist Fernandes (2012) says when he asserts that the food consumption patterns of a domestic group must be perfectly coordinated in order to obtain the greatest caloric and economic return from its resources at each of the different points in the annual cycle, while always seeking to maintain the necessary food diversification throughout this cycle.

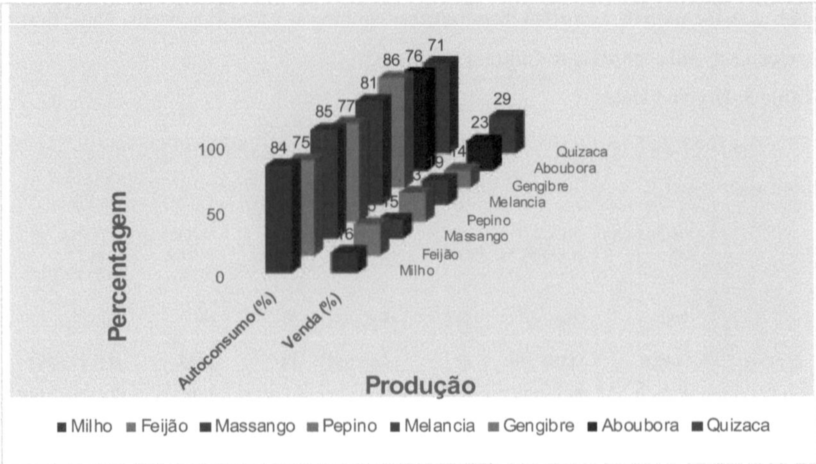

Figure 2. Crops produced under rainfed conditions.

Continuing his idea, he says that the bulk of the region's population's diet is based on milk and the reserves of cereals, particularly maize, millet and millet, that they manage to accumulate. From these products, cornmeal is obtained (cooked with water, without salt, ending in a consistent, soft dough), which can be accompanied by lombis of various branches, beans, almonds, plant extracts and butter, gindungo and tomatoes are added and it can be garnished with dried or fresh fish, mafuma, larvae, meat, etc.

There are around 50,000 species of plants on Earth, 3,000 of which have been used for food, but only thirty crops are responsible for providing 95% of the energy supplied by the food consumed by humans; most of them (60%) are rice, wheat, corn, millet and sorghum (Alisson, 2013).

Another of Fernandes' (2012) criteria is that reasons such as ideology, religion, opinions, habits, moral conceptions, beliefs, traditional laws, etc. influence Angolans' culture of preserving the environment, which has increasingly deteriorated due to the constant intervention of different cultures in the country throughout history. The traditional practices of mono-cropping and post-harvesting are often practiced as a belief and habit of thanking nature for the results of production.

Table 4 shows the total production of horticultural crops by five producers. As can be seen, the crop with the highest production is tomatoes, at 2914 kg, and this is the one that increases total sales revenue the most. Unlike rainfed products, vegetables are used more for sale, in this case 87 % of total production, than for consumption, where only 13 %

was used for family self-consumption; even so, we believe that the profit of 315794 Kz is not enough to meet the needs of the numerous families under study. These results are linked to the lack of permanent water sources and irrigation systems to guarantee this activity. Of the five producers, only one has a hosepipe.

Table 4. Vegetable production

PRODUCTS	USE						
	PRODUCTION	SELF-CONSUMPTION		FOR SALE		SALE PRICE	TOTAL SALES REVENUE
	(KG)	KG	%	KG	%	KZ	KZ
Lettuce	313	68	22	244	78	100	24488
Onions	496	137	28	359	72	200	71800
Carrot	180	41	23	139	77	100	13900
Cabbages	220	65	30	154	70	100	15400
Peppers	98	20	20	80	80	200	16000
Tomato	2914	222	8	2691	92	500	1345500
Eggplant	166	46.76	28	119.24	72	200	39746
TOTAL	**4388**	**600**	**13**	**3788**	**87**	**1400**	**315794**

In addition to these results, we believe that vegetable production is varied, bearing in mind that the research was carried out between January and April this year, and the region has a semi-arid climate and sandy soils, lacking in organic matter, but families could count on some of these products for the whole year if they followed food preservation practices.

These results coincide with those of Fernandes (2012), who in his research classifies the times of year for the southern region of Angola, in which he states that the Hunger Season (scarcity) is from November to January, the Intermediate Season from February to May (the first agricultural products begin to appear) and the Season of Plenty from June to October (abundant food of agricultural origin and the slaughter of livestock).

In similar research, Lundahl et al. (2012) point out that the climate in East Timor is a typical monsoon climate, with a rainy season extending from around December to March and a dry season from May to October, but with local geographical variations. Annual rainfall varies between 500 and 1,500 mm in the relatively dry north and over 2,000 mm in the relatively humid south. Local variations in rainfall are common, since Timor-Leste is ecologically diverse as a relatively small country.

The implications of climate change for food security are notable, especially in the Pacific countries. The study points out that changes in temperature and precipitation regimes (around 70% of the Pacific's agricultural area is heavily dependent on seasonal rainfall), changes in average and peak temperature, loss of genetic resources and agrobiodiversity, increased salinization, more intense climate-related natural disasters, and changes in disease and pest regimes could have significant repercussions for agricultural production (Government of Timor-Leste, 2018).

Costa (2010) states in his research that the average quantities consumed for Vegetables and Legumes were as follows: onions (29.1%), tomatoes (27.9%), carrots (33.) and lettuce (30.9%).

Livestock production is shown in the following table, with a total animal population of 351 head, the largest number of which are cattle (38 %), which are kept almost entirely for years, as they are family assets, followed by goats (28 %), poultry (26 %), pigs (8 %), and finally donkeys (3 %), which are used for transportation, animal traction and loading goods.

These results are representative of the culture that characterizes the peoples of southern Angola, which has been described by yet another researcher.

Table 5. Livestock production

PRODUCTS	USE						
	CURRENT STAFF	SELF-CONSUMPTION		HEADS SOLD		SALE PRICE	TOTAL SALES REVENUE
	U	U	%	U	%	Kz/U	Kz
Cattle	133	125	94	8	6	123.000	984.000
Goats	99	91	92	8	8	35.500	284.000

Birds	90	75	83	15	17	675.0	10.125
Pigs	29	24	83	5	17	15.540	77.700
Donkeys	12	12	100	-	-	-	-
TOTAL	**351**	**315**	**90**	**36**	**10**	**34943**	**1.355.825**

Fernandes (2012) states that in the case of meat consumption, animals (cattle) are slaughtered for consumption not out of preference, but only in extreme cases, due to illness, accidents or advanced weakness. The animals slaughtered for consumption are small ruminants (sheep, goats and pigs), according to the preference of the ethnolinguistic groups. Chickens can also be used for food, but their small size is insufficient to meet the needs of the domestic group, which greatly restricts their consumption.

In studies carried out in the province of Huila by Costa (2010), **the** average quantities consumed for Eggs, Meat and Fish were the reference, in the following percentages, for eggs (), beef, pork and goat (), chicken (43%), ham (), sausages (36%) and fish (). The foods consumed in the highest percentage, in quantities equal to the reference portion, were eggs and beef, pork and goat. The food consumed in the highest percentage, in a smaller quantity than the reference portion, was dried fish. The foods consumed in the highest percentage, in quantities greater than the reference portion, were eggs and beef, pork and goat.

In the OXFAM4 food index, Angola is listed in position 123, meaning that it is one of the countries with the worst food situation. The index evaluates data provided by the FAO and WHO on what is good to eat, the price of food, water quality, diabetes and obesity, among other variables. Some of the organization's explanations reveal that Angola's problem is not a lack of food, but the price of food and the diet of families, especially the nutrition of children (INE, 2020).

Following this, the same source reveals that people spend an average of more than three months with insufficient food. Food insufficiency is a serious social problem that deepens hunger in households and malnutrition in children, having a serious impact on the physical, emotional and social development of those affected, as well as on the income of economically active people. The issue of insufficiency should not be confused with the lack of food or the food crisis. The challenge for families is to obtain enough food to satisfy the number of household members within 30 days.

In the Brazilian semi-arid region, as in other developing regions, animal production stands out for its importance in food security, job and income generation, traction, transportation, and the production of fertilizer and fiber (Lima, 2006). Meat, eggs, milk and dairy products, together with plant products, can meet the nutritional requirements of human beings, as they are sources of lipids, proteins and carbohydrates (Torres et al., 2000).
These animal production systems are related to production capacity, management, social benefits and even the survival of families in situations such as the periodic droughts that affect the Northeast of Brazil. Thus, the diversification of production is generally related to financial conditions, the personal characteristics of the producer and the factors of animal production in the family farming economy, such as periods of water scarcity, soil aridity, distance from large input supply centers, among others (Bezerra et al., 2013).

Existence and manifestations of diseases associated with malnutrition

In evaluating this indicator, I took into account the criteria given by the local public health authorities and observation, which from a qualitative point of view express the existence of conditions such as: rickets, high blood pressure, anemia due to lack of iron, decreased growth, respiratory disorders, immunodepression, diabetes, premature births, low birth weight, among others.

These issues are confirmed by the FAO (2000), which points out that the problems caused by ingesting an inadequate number of calories in relation to what is recommended for each individual (an average of 2100 calories a day) are varied and affect the person's development in a direct and visible way, weakening the immune system and making the individual, especially the child, more vulnerable. Malnutrition, obesity are some of the situations caused by an inadequate supply of calories to the body.

The organization also believes that malnutrition means a situation in which the diet does not include a sufficient number of calories to meet the minimum standard that is physiologically necessary for the body to function properly. The author ends by defining malnutrition as the state in which an individual's physiological functions are so impaired that it is not possible to maintain the body's normal capacities, such as growth, pregnancy, lactation, learning, physical work, endurance and recovery from health problems.
In our opinion, the impact of a lack of food on the family is felt both personally and socially, with consequences ranging from the onset of disease to public safety and economic productivity.

In a study carried out by Simão (2008) in the city of Lubango to study the prevalence of arterial hypertension and risk factors, several aspects associated with malnutrition were identified (excess weight, obesity, high consumption of salty foods); 60.6% of the participants in the study indicated a preference for consuming very salty foods and consumed meals in which carbohydrates and proteins predominated and a low consumption of fruit and vegetables. These unbalanced diets lead to problems of malnutrition which constitute risks associated with the development of diseases.

Representing serious public health problems, malnutrition contributes very significantly to reduced cognitive and learning abilities, school dropout, poor physical and emotional development in children and a life more susceptible to infections and low productivity rates during working life. In fact, the World Health Organization estimates that malnutrition is linked to one third of child deaths (WHO, 2018).

Technical training in the region

With regard to technical training for the producers in the selected sample, only of the producers said that they sometimes received technical training. 67% said that they had never received technical training from the professionals in charge of the task. This makes it very clear that there is a need to extend training to these households, in order to improve production results and change household food insecurity.

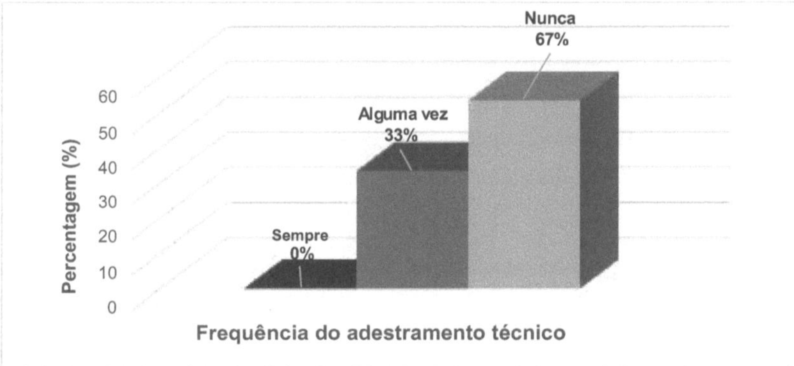

Figure 3. Professional technical dressage

According to the results of this indicator, there is insufficient training for livestock producers and breeders, which influences productivity and poor feed management. This coincides with the criteria of Rendo et al. (2015), who state that technical training is a process of dynamic transformation of accumulative change in local rural societies, which, with the participation of social actors, makes it possible to diversify both productive and income-generating activities, as well as forms of social organization and political participation, and to achieve technological development in the midst of cultural diversity.

According to Landini (2015), in the results of his research "The conception of rural extension at a qualitative level", he mentions that the actions involved, such as working with groups of producers, strengthening links and managing conflicts, (60.7%) agree that in this line "the organization of activities in an associated way" and "the application of group methodologies" as a fundamental element makes sense if we take into account that in Argentina most of the actions of Rural Extension are oriented towards working with groups of producers.

In this sense, Huesca et al. (2019), in their research, found that 79% of producers rated the process of communication with producers by the technicians in the Povoa Plan as excellent, while only 2% of those interviewed considered MasAgro's extension activities to be excellent.

The preservation of both traditional knowledge and native and exotic vegetation introduced in the past, which are used by populations, is necessary, as the rescue of potentially useful species is of fundamental importance for the maintenance of local eco-systemic and social processes (Santilli, 2009). According to Amorozo (2007), this process of interference in the maintenance of traditional knowledge has limited its flexibility and adaptive capacity. According to the author, reasons such as the overcoming of use value by market value, the introduction of more aggressive exploitation technologies, changes in population density, local subsistence conditions, unemployment, among others, have led to changes that are difficult to reverse, both for the environment and for the populations linked to it.

Locally constructed knowledge associated with natural resources has been threatened by extreme changes, far beyond its resilience potential. The erosion of this knowledge is closely associated with the abandonment of resource management schemes and changes in people's way of life (Berkes, 1999).

4.2 Proposal for rural extension work to promote food security in the locality of Ovipembe, Ondjiva commune.

The fundamental aim of these actions is to provide education and technical assistance for the introduction and expansion of sustainable and resilient agriculture, the improvement of organic and family farming and soil and water conservation in agricultural systems.

This proposal is based on the principles of rural extension, as a dynamic and participatory activity, adapted to the ethnic, cultural, sociological and economic realities of the area in question

It is made up of 5 actions, which respond to the most pressing needs of the population under study, but it is considered that other themes will be necessary according to the reality of the region and food security.

Action 1: Sustainable family farming

Objective: To promote the sustainable development of agriculture as a secure source of food production.

Methodology: to spread the word about the importance of agro-ecological production systems through rural extension, focusing on organic production and the process of converting to it. Access to the activities is free for all producers and their families in the region.

Content: Benefit of organic production for the population, seeking to create direct social, environmental and commercial links in rural locations. Sustainable production practices.

Mediums: promotional materials on family farming and organic production. Videos, photographs reflecting experiences from other regions.

Practical activity: Training and orientation of farming families (including young people and women) in production issues and recognition of the use of family labor in production areas, through lectures, promotion and participation in events with stands and other forms of group and participatory activities in production areas.

Goals

- Increase agricultural productivity, ensure sustainable food production systems and maintain genetic diversity.
- Increasing the quantity and variety of its products.
- Increasing the production and consumption of organic food

Action 2: Soil and water conservation in agricultural ecosystems

Objective: to describe and adopt sustainable agricultural production systems as a fundamental element in meeting the family's growing demand for food.

Content: soil and water management practices and their increase in productivity. Soil and water conservation activities in their area of influence, to reduce the introduction of sediments, nutrients and pesticides into tributaries, rivers and reservoirs.

Practical activity. Demonstration of soil conservation activities. Preparation and application of organic and chemical fertilizers. Calculating doses when applying chemical fertilizers.

Goals

- Implementation and use of good soil management practices, such as the no-till system, permanent soil cover, crop consociation and crop rotation and diversification.
- Adoption of safety measures to avoid environmental contamination.

Action 3: Marketing surplus production inside and outside the communities.

Objective: To explain how to expand marketing not only in the communities in which they are produced and with greater added value to the products, thus providing improved income and economic development.

Content: technical and economic aspects in the development and continuity of family farming. Production and marketing of fruit seedlings, medicinal plants, seeds, among others.

Medium: stands and promotional materials on labelling, nutritional information and product standardization and the use of medicinal and phytotherapeutic plants.

Practical activity: Drawing up nutritional factsheets on natural products and their common uses.

Goals

- Increasing the levels of commercialization of the products of the mines and their transformation.
- Increasing the quantity and quality of products on offer (ornamental, medicinal and other plants)

Action 4: Production of animal feed.

Objective: To create forage support by establishing areas for the production of native pastures, with productive, protein-rich and drought-resistant crops.

Contents. Protein availability through the production of essentially exotic legumes. Creation of protein banks. Fodder conservation techniques to guarantee the animals' food supply throughout the year.

Practical activity. Identifying the local potential for producing abundant green mass to provide fodder (grasses and legumes). Demonstration of making hay and silage as alternative food preservation techniques for the dry season.

Goals

- Increase livestock productivity by implementing sustainable animal food production systems.
- Increasing the quantity and variety of meat and milk by-products.

Action 5 education

Objective: to promote the use of healthy recipes, based on the principles of the Food Guide drawn up by Costa (2010), in order to learn about the quality and quantity of food needed by the body.

Content: food safety, healthy eating, food reuse and environmental conservation.

Practical activity: demonstration of the conservation and processing of animal and vegetable products. Preparation of typical local dishes using the new nutritional knowledge.

Goals

- Increase consumption of organic and nutritious food without losing culture and traditions.

Expected results

With these training actions, it is clear that the connection between modern agriculture and sustainable agricultural practices is very important for increasing food, reducing hunger and improving the nutrition of families. Basic knowledge will also be acquired on the subject of soil and water, cover crop seed production, crop rotation and diversification, invasive plant control, production system management, pest management and diseases, animal feed, among others. Based on this knowledge, actions to encourage and favor greater diversification in the use of food plants not only as an economic alternative for rural producers, but also broaden specific knowledge about such plants and local products.

V. CONCLUSIONS

1. Even when the producers under study own land and practice traditional agriculture and livestock farming, their production levels are insufficient to guarantee food security for their families.
2. Livestock farming in the town is in keeping with the customs and traditions of the region, where cattle are the family's heritage, and goats, pigs and poultry are used to a lesser extent to obtain cash in exchange for buying other basic products in the town's stores and stores.
3. There are manifestations of diseases that respond to malnutrition, mainly in school-age children and pregnant women.
4. A proposal was drawn up for rural extension work to promote food security in the locality of Ovipembe, Ondjiva commune.

VI. RECOMMENDATIONS

1. Carry out studies related to the role of rural extension and its influence on food security in the locality under study and experiences in others.
2. To disseminate the results obtained in this work, in the educational centers of the Ondjiva commune and other localities in the province of Cunene, with the participation of competent institutions.

VII. BIBLIOGRAPHICAL REFERENCES

1. Alisson, E. (2013). *Biodiversity loss is a global problem*. Agência FAPESP, 2013. Available at: http://agencia.fapesp.br/17544.
2. Altieri, M. A. & Nicholls, C. I. (2009). *Agroecological challenges for the development of sustainable agriculture in the 21st century*. Conference presented at, Salón del Capitolio Nacional. La Habana. Cuba.
3. Altieri, M. A. (1997). *Agroecology. Scientific support for sustainable agriculture. Clade-ACAO*. Lima, Peru. Editorial Nordan-Comunidad. pp 249.
4. Amorozo, M. C. M. (2014). *Traditional agricultural systems and the conservation of agrobiodiversity*. 2002. Available at: http//:www.ambiente.sp.gov.br/ea/adm/admmarqs/MariaA.pdf.
5. Andrade, G. D. Santos and Chagas, R. de C. S. (2012). *Technical assistance and rural extension at the agricultural family school in the sertão: between the needs of peasants and the MDA's ATER policy*. São Paulo.
6. Angola (2005). *Millennium Development Goals Ministry of Planning Largo do Palácio, Cidade Alta, Luanda*. UNDP. Available at: http://mirror.undp/org./angola/publications.htm.
7. Bellón, M. (2002). *Métodos de investigación participativa para evaluar tecnología: manual para científicos que trabajan con agricultores* (p.106). México, D.F.
8. Berkes, F. (1999). *Sacred Ecology: traditional ecological knowledge and resource management. Philadelphia*: Taylor & Francis, 1999.
9. Bettiol, W., Ghini, R., Montes, J., Haddad, A. & Siloto, R. C. (2014). *Organic and conventional tomato cropping systems. Sci. Agric*. 61(3): pp253 - pp259.
10. Burton, C. (1991). *Extension for Development*. Brazil
11. Carvalho, M. (2003). *What is nature*? Editora Brasiliense: Colecção Primeiros Passos. 2. ed. São Paulo,
12. Castro, J. (1968). *Geopolitics of Hunger: an essay on the problems of food and population*. 1st volume. Editora Brasiliense. São Paulo. 1968.
13. CEIC/UCAN. (2021). *ANGOLA ANNUAL ECONOMIC REPORT 2019*, Catholic University of Angola, Luanda.
14. Costa, P. C. (2010). *Characterization of Food Habits in Huíla Province, Angola*. Dissertation submitted as a partial requirement for the degree of Master in African

Studies: Analysis and Management of Social and Economic Development, Center for African Studies, ISCTE-IUL October, 2010

15. Cultrera, M. (2008). *Ethnobotanical study of food plants cultivated by residents of the outskirts of Santo Antônio de Leverger, MT* (dissertation). UNESP, 2008.
16. Dias, C. J. (2015). *Food security: Actions promoted by the non-governmental organization VIDA - Volunteering for African Development.* Varela, Cacheu Region, Guinea-Bissau, 2015. © Fernando Mendes.
17. Engel, P. (2000). *Facilitando el desarrollo sostenible: ¿hacia una extensión moderna?* Centro de Estudios y Gestión para el Desarrollo Rural Sostenible (CEDRO), Universidad de Concepción, Chile.
18. Engel, P. (2000). *Facilitando el desarrollo sostenible: ¿hacia una extensión moderna?* Centro de Estudios y Gestión para el Desarrollo Rural Sostenible (CEDRO), Universidad de Concepción, Chile. Mimeo.
19. Espinoza, G. (2001). *Fundamentals of Environmental Impact Assessment*. IDB - CED. Santiago de Chile. CHILE. 186p.
20. FAO (2020). *State of Food Insecurity and Nutrition in the World 2020 (SOFI).* Food and Agriculture Organization of the United Nations, Rome.
21. FAO (Food and Agriculture Organization of the United Nations) (2014). *The State of Food Insecurity in the World.* 2014, SUMMARY.
22. FAO. (1996). *World Food Summit.* Rome. http://www.fao.org/docrep/003/w3613p/w3613p00.htm
23. FAO. (2000). available at: http//:www.fao.org.
24. FAO. (2003). *Livelihood Support Program* (LSP). Rome. ieeh faazrgheide-es.html.locale.es. Consultation date: January 2, 2021.
25. FAO. (2014). *The state of food and nutrition security in Brazil - A multidimensional portrait* - 2014 report. FAO.
26. FAO. (2015). *Agricultural Production, primary crop* (Report #1). Available at: hfaazrgheide-es.html.locale.es.
27. Fernandes, G. (2012). *Livestock in family farming in southeast Angola, livelihoods and vulnerability of pastoralist and agropastoralist household groups.* Technical University of Lisbon, Higher Institute of Agronomy. Lisbon.
28. Ferrão, J. E. M. (2002). *Concepts of Food Security. Consequences in the poorest countries*. Gazeta das Aldeias. Nº 3094: 33-39

29. Fortes, M. M. F. (2012). *FOOD SECURITY IN CAPE VERDE. Case Study in the Municipality of Ribeira Grande, Santo Antão Island.* Dissertation for a Master's Degree in Tropical Agronomy and Sustainable Development. Lisbon 2012
30. Freire, A. G., Melo, M. N., Silva, F. S., and Silva, E. (2015). *Around the house, the animals of the yard.* Agriculturas, Rio de Janeiro, v. 2, n. 4, p. 20-23.
31. Government of Timor-Leste (2017). *Enactment of the Land Laws is a decisive step towards development.* Available at: < http://timor-leste.gov.tl/?p=18216 >.
32. Huesca, J.M., Hernández, M., Hernández, O., Fernández, Y., Díaz, H., & Estrella, N. (2019). *Extensionism in regional agricultural programs: Plan Puebla and MasAgro.* Revista de Alimentación Contemporánea y Desarrollo Regional, Volumen 29 (Número 53), Electronic journal. ISSN: 2395-9169.
33. Huesca, J.M., Hernández, M., Hernández, O., Fernández, Y., Díaz, H., & Estrella, N. (2019). *Extensionism in regional agricultural programs: Plan Puebla and MasAgro. Revista de Alimentación Contemporánea y Desarrollo Regional,* Volumen 29 (Número 53), Electronic journal. ISSN: 2395-9169.
34. INE (National Statistics Institute) (2020). *Thematic report on gender.* Expenditure, income and employment survey in Angola 2018-2019. Luanda - Angola. Available at: www.ine.gov.ao
35. INE, (2018). *Inquérito Integrado sobre o Bem-estar da População - 2010,* Ministry of Planning, Luanda.
36. INE. (National Statistics Institute) (2022). *Economy and finance. Wholesale price index* Luanda, Angola - 2022. Available at : geral@ine.gov.ao
37. INE. (National Statistics Institute) (2020). *Thematic report on gender 2018-2019* Luanda - Angola 2020. Available at: www.ine.gov.ao
38. Jacobsen, B.H., Pedersen, D.E., Christtensen, J., Rasmussen, S. (1994). *Farmer's decision-making - a descriptive approach.* Proceedings from the 38th EAAE Seminar 1994, October, Kopenhagen, p. 77-89.
39. Jiménez, W. (2015). *Training guide on agricultural issues for family farmers, United Nations Food and Agriculture Organization, Managua.*
40. Jonsson, U. (1981). *The causes of hunger. Hunger and malnutrition, social determinants/Flávio* Luiz Valente (org). - São Paulo: Cortez, 1986.

41. Lacki, P. and Marcenaro, L. (2005). *A new training for rural development ¿Spending on activities or investing in results? ¿Problematizing solutions or solving problems?*
42. Landini, F. (2015). *Concepción de "extensión rural" de los extensionistas rurales argentinos que trabajan en el sistema público nacional con pequeños productores.* Available online: 2015-15-03 doi:10.11144/Javeriana.cdr12-75.cere. Date of consultation 5/01 2021.
43. Landini, F. and Bianqui, V. (2014). *Rural extension in Ecuador from the point of view of its extensionists.* Revista de la Facultad de Agronomía, 31(3).
44. Leyva, R. L., Baldaquín, P. A., Ruz, R. R., Ayala, J. R. & Pupo, F. (2010). *Participación campesina en el diagnóstico de la calidad de los suelos en fincas agropecuarias de la provincia de Las Tunas. DELOS: Desarrollo Local Sostenible, 9(*3).
45. Lima, G.F. da C. (2006). *Alternatives for the production and conservation of strategic forage resources in the northeastern semi-arid* region. In: ENCONTRO NACIONAL DE PRODUÇÃO DE CAPRINOS E OVINOS, 1, 2006, Campina Grande. [Proceedings]. Campina Grande: SEDAP, 2006.
46. Loliana Sive, de O., Ilunga, K. R., and Massochi, I. P. (2021). *Final report on the survey and characterization of local markets.* October 2021
47. Lopez, R. C. (2007). Sistemas de producción agrícola sostenible en los Andes de Venezuela: Agricultura. Avances de química, 23-33. doi: 1856-5301
48. Lundahl, M., Sjöholm, F. (2012). *Improving the Lot of the Farmer: Development Challenges in Timor-Leste during the Second Decade of Independence.* IFN Working Paper No. 929. 2012. Available at: < http://www.ifn.se/wfiles/wp/wp929.pdf>. Accessed on: June 3, 2018.
49. Magalhães, F. (2021). *Development and Food and Nutrition Security: Perspectives and (un)balances.* FEC - Faith and Cooperation Foundation: Lisbon.
50. Magalhães, F. P. (2018). *Food and Nutrition Security and Development.* June 2018 legal deposit copyright © FEC - Fundação Fé e Cooperação and IMVF - Instituto Marquês de Valle Flôr
51. Mendonça, L. K. and Galindo, E. (2015). *Integration of social and rural development policies for productive inclusion and overcoming extreme poverty in the Brazil Without Poverty Plan.* In: MELLO, J. A inclusão produtiva rural no

Brasil Sem Miséria: o desafio da superação da pobreza no campo. Brasília: MDS. (p. 32-48). (Cadernos de Estudos Desenvolvimento Social em Debate, no. 23).

52. Millennium Development Goals (MDG) (2010). Available at: http://www.un.org/spanish/millenniumgoals/pdf/MDG_Report_2010_SP.pdf.
53. MINADER. (2004). *National Medium-Term Investment Program under the PDDAA-NEPAD (PNIMP).*
54. MINADER. (2020). *2019/20 Agricultural Campaign Report*
55. MINADER. (2022). *2020/21 Agricultural Campaign Report*
56. MINAGRI. (2019). *2017/18 Agricultural Campaign Report*
57. Münch, L. (2005). *Organización: diseño de organizaciones de alto rendimiento*; Mexico; trillas.
58. WHO, (2008). *Cit in "Nutrition Sector Analysis of Children and Women in Angola"*. Boukare, V. et al. UNICEF
59. Ortiz, T. A. and Astier, C. M. (2003). *Systematization of agroecological experiences in Latin America. LEISA Journal of Agroecology*. Special issue.
60. OXFAM SITE: https://www.oxfam.org.uk/what-we-do/good-enough-to-eat
61. Ramírez de Haro, G. & A. Rodríguez C. (2002). *"Instituciones y desarrollo rural" en Desarrollo y cooperación en zonas rurales de América Latina y África*.38pp.
62. Rendón, R., Roldán, E., Hernández, B. & Cadena, P. (2015). *The processes of rural extension in Mexico*. Revista Mexicana de ciencias agrícolas, 6(1), 151-161.
63. Rendón, R., Roldán, E., Hernández, B. & Cadena, P. (2015). *The processes of rural extension in* Mexico. Revista Mexicana de ciencias agrícolas, 6(1), 151-161.
64. Republic of Angola, Ministry of Agriculture and Fisheries and National Directorate of Agriculture and Livestock (2021). *Situation of Acute Food and Nutrition Insecurity in the Provinces of Cunene, Huila and Namibe*. Luanda, August 2021
65. Republic of Angola (2009). *National Strategy for Food and Nutrition Security* (ENSAN) Luanda, March 2009
66. Republic of Angola (2020). *Fundamental Report of the General State Budget* (OGE 2020).
67. Riches, G. (2002). *Food banks and food security: Welfare reform, human rights, and social policy*. Lessons from Canada? Social Policy & Administration, 36 (6), 648-663.

68. Roa, S., Gómez, N., Lozada, B., and Vargas, A. (2002). *Identifying priority problems in rural communities through participatory diagnostics* Geoenseñanza, vol. 7, núm. 1-2, 2002, pp. 30-37 Universidad de los Andes San Cristobal, Venezuela

69. Rocha, C., Mora, J. and Romero, J. (2016). *Typology of production systems in the rural area of the municipality of Ibagué*, Colombia. Agronomía Mesoamericana 27(2), 253-264. doi: 10.15517/am.v27i2.2436

70. Rodríguez, M. (2004). *Desarrollo endógeno local de productores rurales de chile poblano (Capsicum annuum L.) en dos comunidades rurales de Puebla, México.* Thesis presented as a partial requirement for obtaining the degree of Doctor of Science. Postgraduate College, Institution for teaching and research in agricultural sciences. Campus Puebla. pp 22- 30.

71. Ruíz, J. (2009). *Application of bioproducts to ecological tomato production.* Cultivos Tropicales 30(3): pp 60 - pp 64.

72. Saavedra, G. (2013). *Introduction to vegetable production. Technical Series: Producción de hortalizas para la República de Guinea Ecuatorial.* Number 1. Food and agriculture organization of the United Nations (FAO). Subregional Office for Central Africa. Republic of Guinea Ecuatorial.

73. Santilli, J. F. R. (2009). *Agrobiodiversity and farmers' rights* (thesis). Pontifical Catholic University of Paraná, 2009.

74. Sarandón, J.S., Zuluaga, S.M., Cieza, R; Gómez, C., Janjetic, L. y Negrete E. (2006). *Evaluación de la sostenibilidad de sistemas agrícolas de fincas en Misiones, Argentina, mediante el uso de indicadores.* Revista de Agroecología, vol 1, p 19-28.

75. Scialabba, N. (2011). *E-H. FOOD AVAILABILITY AND NATURAL RESOURCE USE.* 2011. Available at :< http://www.fao.org/fileadmin/user_upload/suistainability/Presentations/Availability.pd

76. Simão, M., Hayasgida, M., Santos, C.B., Cesarino, E.J., Nogueira, M. S. (2008). *La Hipertension Arterial entre Universitarios de la Ciudad de Lubango, Angola*, Rev Latino-Am Enfermagem, July-August; 16 (4), available at: www.eepr.usp.br/rlae

77. Simões, J. (2019). *Commercialization as a factor of social promotion in Angola*, Edição do Fundo de Comercialização, Luanda.

78. Souza, R. J., Fontanetti, A., Fiorini, C. V. A. and Almeida, K. (2003). *Growing beet: conventional and organic cultivation*. Lavras: UFLA,. 37 p.

79. Swanson, B. (1991). *History and evolution of Rural Extension; in Rural Extension: Reference Manual, 2nd ed.* Food and Agriculture Organization of the United Nations, FAO.

80. Torres, E.A.F.S., Campos, N.C., Duarte, M., Garbelotti, M.L., Philippi, S.T. and Minazzi-Rodrigues, R.S. (2000). *Centesimal composition and caloric value of foods of animal origin.* Ciência e Tecnologia de Alimentos, v.20, p.145-150, 2000. DOI: 10.1590/ S0101-20612000000200003.

81. Vega-Dorantes, M. G., & López-Barbosa, L. A. (2014). *The New Rural Extensionism in Mexico.* Universidad Autónoma Agraria Antonio Narro. 65 pp.

82. Zaslavsky, N. (2014). *Rural extension in the no-till vegetable system.* Federal University of Santa Catarina. Center for Agricultural Sciences. Agronomy degree course. Florianópolis. Nov.14-22pp

VIII. ANNEXES

Annex 1. Survey of farmers in the locality of Ovipembe, Ondjiva commune.

As part of the final assignment for the Agronomy Engineering course, a study is being carried out in the locality of Ovipembe. The data collected is confidential and is intended to be the result of research into the population's food security. Thank you very much for your participation.

1. Socio-demographic characterization

- No. of people in the household: H___ M___ Total___

Age group	0-14		15-19		20-59		> 60		CA schooling and complementary 1⁰				
	H	M	H	M	H	M	H	M	Can't read	You can read	Prim.	Sec.	Sup.
Head of Household CA)													
1⁰Complementary													

2. Characterization of Agricultural Activity

Tranche	Area (m^2 or ha)	Form of Operation	Paid Income (in Kz) (If applicable)
Sequeiros			
Irrigation			

2.1 Production of dried fruit and destination of production (Estimates for the last 12 months)

Products	Use				
	Production (Kg)	**Self-consumption (%)**	**Selling (%)**	**Sales price (Kz)**	**Total revenue Sales (Kz)**
Corn					
Beans					
Sweet potatoes					
French fries					
Cassava					
Yam					
Vegetables					
Fruit trees					
Other (please specify)					

2.2 Horticultural Production and Destination of Production (Estimate for the last 12 months)

Main vegetables and fruits	Use				
	Production (Kg)	Self-consumption (%)	Selling (%)	Sales price (Kz)	Total sales revenue (Kz)
Pumpkin					
Lettuce					
Reindeer potatoes					
Sweet potatoes					
Onions					
Carrot					
Cabbages					
Beans					
Peppers					
Carrot					
Cabbage					
Tomato					
Banana					
Mango					
Papaya					
Other (please specify)					

2.3 Sale and marketing of vegetables

2.3.1 How vegetables are sold

- At home ______________
- Ambulante ____________
- On the market ____________
- Other (please specify) ____________________________

2.3.2 Main marketing constraints?

----------------.

3. Livestock production and destination of production (estimate for the last 12 months)

Type of animal	Effective Current	Self-consumption	Heads Sold	Sales price (Kz/unit)	Total sales revenue (Kz)

Annex 2. Methodology used to design training actions, according to Jiménez (2015)

The overall aim is to provide a technical-methodological tool for developing training for family farmers on agricultural issues, using participatory techniques that facilitate the participants' learning.

The methodology described is participatory. It suggests the development of group sessions, reflections, presentation dynamics, practical work and evaluation, with the aim of energizing and strengthening the learning process and encouraging new new attitudes and values such as initiative, creativity and discipline. The activities can be carried out in the classroom and in the field, linking theory with practice.

The proposal is structured in such a way that it can easily guide the facilitator towards achieving the objectives, it includes topics that have been identified as a priority for the technical development of family farmers and the topics covered include a method sheet with the following information which includes the following information:

- Objectives;
- Duration of the activity;
- Methodology;
- Materials to be used in the execution, both theoretical and practical (you can replace some of the suggested materials with those you have available at the time);
- Content development.

  General considerations that will allow the facilitator to obtain a better understanding and fixation of the participants' learning process:

- Before starting the activity, the facilitator should find out the type of participants and their level of education. This will enable them to prepare the material to be used in the training;
- Avoid giving very long theoretical debates by planning participatory activities such as group work, visits to plots, etc;
- Avoid using very technical or scientific concepts that limit the understanding and learning of family farmers;
- Methodologically, the group should not exceed 30 participants. Each participant should receive the necessary working materials to understand the topic, depends

entirely on the technical and methodological suggestions defined in the planning process of the outreach activity.

- Training small groups ensures:

✓ The achievement of its objectives;
✓ The greatest uptake during the learning process;
✓ Fulfillment of agenda times;

- To begin the activity, it is suggested that the participants introduce themselves and explore their expectations, which will facilitate the evaluation of the learning obtained during the training;
- Write down with the participants the golden rules that will prevail during the training sessions, for example: group support, rules of active participation, respect for colleagues, respect for the schedule of activities, punctuality, among others that the participants may suggest;
- For group activities or visits to field plots, it is suggested to form groups with a maximum of 6 participants, with the aim of facilitating active participation and improving the learning process.

Annex 3. Agricultural production system with irrigation

Annex 4. Dryland farming

Annex 5. Livestock farming

MIX
Papier aus verantwortungsvollen Quellen
Paper from responsible sources
FSC® C105338

Printed by Books on Demand GmbH, Norderstedt / Germany